If You Give An Officer A Spreadsheet

by

W.G. Reynolds

illustrated by

Ichigodawashi

This page intentionally left blank

If you give an officer a spreadsheet,

he will ask for it in color.

If you give it to him in color,

he will ask for a briefing on it.

Once you are prepared to give the briefing,

he will ask for a Paw-R-Point of the briefing and he will want it projected on the big screen.

If you project the briefing on the big screen,

he will like how it looks and will ask you to invite all of his officer friends so they can see it too.

If you invite his officer friends to see the briefing,

he will want to give the briefing himself and will need a room big enough for all of them.

Therefore, he is going to ask you to reserve the base theater.

When you reserve the base theater,

he is going to want snacks for all his friends and will ask you to get some.

If you get snacks,

he will become conscious of his image and he will want more than just snacks. He will ask you to have the briefing catered.

If you get the briefing catered,

he will finally decide that it is good enough for him to give the presentation.

All of his officer friends will be deeply impressed.

While he leaves you to clean up the base theater, he will be talking with all of his officer friends when the Good Idea Fairy will strike!

They will have the great idea to work together on a tabletop exercise and he will ask you to plan it.

If you start planning the table top exercise,

he will realize he needs to pay for it and will ask you to put together a finance meeting.

If you put together a finance meeting,

he will find out that he has funds that he needs to spend before the end of the fiscal year and he will ask you to find an exotic location for the table top exercise.

The first location will be too hot.

The second location will be too cold.

The third location will be just right and he will ask you to book an appropriate venue.

The first venue will have chairs that are too hard.

The second venue will have chairs that are too soft.

The third venue will have chairs that are just right and when you report back that you found the right place...

he will ask you to completely scrap the plans for the table top exercise because instead you are going to plan a live fire exercise!

If you start working on the plans for a
live fire exercise,

he will ask you to coordinate range time.

If you coordinate range time,

he will ask you for a weather forecast.

When you give him a weather forecast,

he will ask you to arrange for airlift.

If you arrange for airlift,

he will realize he has extra space on the aircraft and will ask you to coordinate with Research & Development (R&D) so they can bring their newest super-secret firework and participate in the live fire exercise too.

If you coordinate with R&D,

he will become conscious of his image and he will realize that this new firework is going to outdo anything that he has.

Not wanting to be outdone at his own event he will ask you to get the nuclear firework codes.

If you get him the nuclear firework codes,

he will need a report written and will ask you to provide for the report....

a spreadsheet.

Dedicated to
my wife who encouraged this work
&
to all of the veterans and current service men & women.
Remember when things get tough, it is ok to laugh about it.

About the Author & Illustrator

W.G. Reynolds is a technical writer who currently resides in San Diego, California where he writes boring test plans, reports, and contract proposals. Prior to being a technical writer, he was a network engineer who installed comm equipment on Japanese warships, and is a 16-year veteran of the United States Air Force (USAF) as a Tech Controller (3C2X1) to a combat communications unit where he would describe it as 'camping with internet'. W.G. Reynolds has a love of ferrets and their roguish behaviors that pairs well with his unique sense of humor.

Ichigodawashi is an accomplished artist that has dedicated her artwork to the animal family of Mustelids and specializing in ferrets. She loves manga, anime and has been drawing since she was a little girl. She developed a love affair of ferrets when she starting living with them in 2018 and has been completely fascinated by their cute and mischievous behaviors. Ichigodawashi currently resides in Japan and has plans to own a ferret art store in the future with hopes to make the world a happier place with her ferret art.

Find more of her artwork on X and Instagram @ichigodawashi

I just got this great idea! I have these pictures and I think they need to be in color. Will you color them for me while I'm out of the office? Don't forget to ask a grown-up for help cutting out the page.

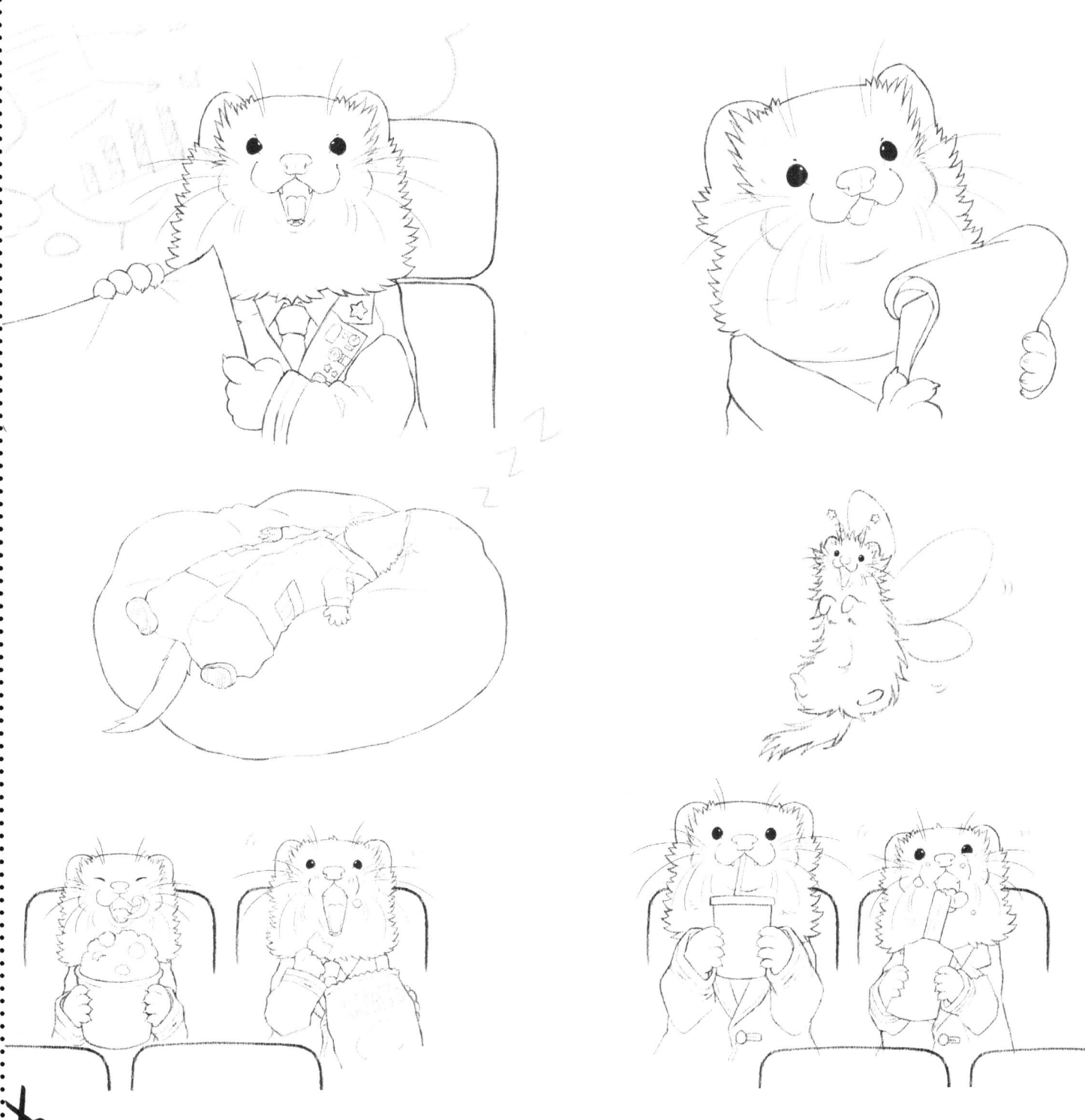